Quilts
from Simple Shapes

RECTANGLES

Mary Coyne Penders

Quilt Digest Press

Editorial and production direction by Harold Nadel.
Book and cover design by Kajun Graphics.
Photography by Sharon Risedorph.
Computer graphics by Kandy Petersen.
Typographical composition by Jet Set.
Printed by Nissha Printing Company, Ltd., Kyoto, Japan.
Color separations by the printer.
Home graciously lent by Margaret Peters.
Construction specifications by Lenore Parham.

Special thanks to Cotton Patch, Lafayette, California;
G Street Fabrics, Rockville, Maryland; and Quilt Patch,
Fairfax, Virginia.

Second Printing

Library of Congress Cataloging-in-Publication Data

Penders, Mary Coyne, 1931-
    Quilts from simple shapes : Rectangles / Mary Coyne Penders.
        p.   cm.
    ISBN 0-913327-35-2 (pbk) : $8.95
    1. Patchwork—Patterns.   2. Quilting—Patterns.   3. Rectangle.
I. Title.
TT835.P448   1991
746.9′ 7—dc20   91-36292
                                                     CIP

The Quilt Digest Press
P.O. Box 1331
Gualala, California 95445

ACKNOWLEDGMENTS

The quilts in this book are alive with vibrant color and visual texture images, thanks to the artistic sensibilities of Alex Anderson. Her incredible energy and enthusiasm sang out through the telephone wires that made our east coast–west coast collaboration possible. Alex's creative talent continues to sing out through the fabrics she combines with flair and exuberance. Always brimming with ideas, Alex infused our working relationship with the joy of making quilts. Alex is indebted to Rosalie Sanders for her encouragement and support, and to Rhondi Hindman for her extraordinary last-minute assistance. My heartfelt thanks to you all.

Lenore Parham tackled the specifications for making each quilt, developing expert guidelines and instructions. Kay Lettau pieced the Special Project with skill and imagination. The Quilt Patch staff contributed frequent advice, assistance and support. I am most grateful to each of you.

To my Quilt Digest Press family, I extend my gratitude for your special skills, dispensed with great generosity: to Harold Nadel for wise editing and production, to Jeff Bartee for knowledgeable direction and support, and to Sharon Gilbert for marketing expertise.

Finally, at hearth and home, my loving thanks to my husband Lee and to my son Christopher, for their encouragement, understanding and myriad helpful services throughout the book gestation process. I am also grateful to my son John for his interest in my work, and the abundant love that accompanies it.

◆ ◆

For Michael, in remembrance of the vision
and inspiration of his leadership and friendship.

INTRODUCTION

I love rectangles! In addition to being easy to work with, rectangles offer opportunities to experiment with color and fabric.

Quilts from Simple Shapes: Rectangles introduces six inviting possibilities, each distinct from the others, made just for you from today's cornucopia of gorgeous fabrics. I know you'll enjoy looking at the variety of different patterns that can be made from this simple shape.

If you're a beginning quilter, this is the ideal book for you. There are six easy-to-make projects in *Quilts from Simple Shapes: Rectangles*. You'll like the simplicity of the easy rectangle shape, which is used for all the designs in this book. Because the template is easy, and the cutting and sewing are easy, you are free to concentrate on choosing colors and fabrics that result in great-looking quilts. I'll help you with lots of color and fabric suggestions.

If you're an experienced quilter, you've probably worked with rectangles in the ever-popular Log Cabin design. Because you're already bitten by the quilting bug, you want to make more and more quilts, but you're frustrated by the problem of too much inspiration and too little time. Have you considered some of the other attractive rectangle patterns which are ideal for quick, easy quilts? In *Quilts from Simple Shapes: Rectangles*, you'll discover quilts for commemorating an anniversary, or enlivening a teenager's bedroom, or making dormitory living more home-like. You can finish a quilt for your baby or your graduate or your grandchild without quitting your job and deserting your family! Best of all, you won't have to sacrifice quality to the pressures of time.

The quilts in this book are also perfect for filling empty wall spaces quickly with lively small quilts. When you study the photographs, you'll probably get lots of ideas for your own home as well as for friends and family. I've thought about beds, walls and tables as I planned the quilts, but I wouldn't be at all surprised to hear that you had thought of even more ways to enjoy quilts. One student called to tell me that her horse loved his new blanket, and another imaginative woman made seat-covers for her Volkswagen!

You can decide if you want to make a quilt exactly as shown in *Quilts from Simple Shapes: Rectangles*, or if you prefer to work with your own color and fabric ideas. When you feel comfortable with rectangles, I hope you'll try different possibilities. I'm sure you'll enjoy the appealing patterns and quick results as much as I have. Once you get started, I bet you'll make them all!

P.S. You'll find a lovely SPECIAL PROJECT in this book, made entirely from rectangles in a Log Cabin design. This is a great gift item that you'll probably want for your own home as well. I promise not to spill anything if you invite me for lunch as soon as you've finished!

Here's what *Quilts from Simple Shapes: Rectangles* offers you:

◇ Hints to help you choose colors and fabrics
◇ Use of your scrap collection
◇ Templates in various sizes
◇ Yardage requirements
◇ Instructions for regular and quick cutting
◇ Directions for construction
◇ Teaching Plan

SUPPLIES

To make the quilts in this book, you will need:

Template plastic

Scissors or knife for cutting template plastic

8″ Fabric scissors or rotary cutter and cutting mat

C-Thru plastic ruler, 2″ x 18″, or quick-cutting ruler

Fine-line pencil

Silver or white pencil for dark fabrics

Pencil sharpener

Glass-head pins

Seam ripper

Sewing machine or hand-sewing needle

Cotton sewing thread

Hand-quilting needle, between #8, #9 or #10

Quilting thread

Pressing surface

Iron

Batting

Quilting hoop or frame

Fabrics: yardage is given for each individual quilt

Containers (such as shoe boxes) for organizing fabric pieces

Good lighting and stable working surface

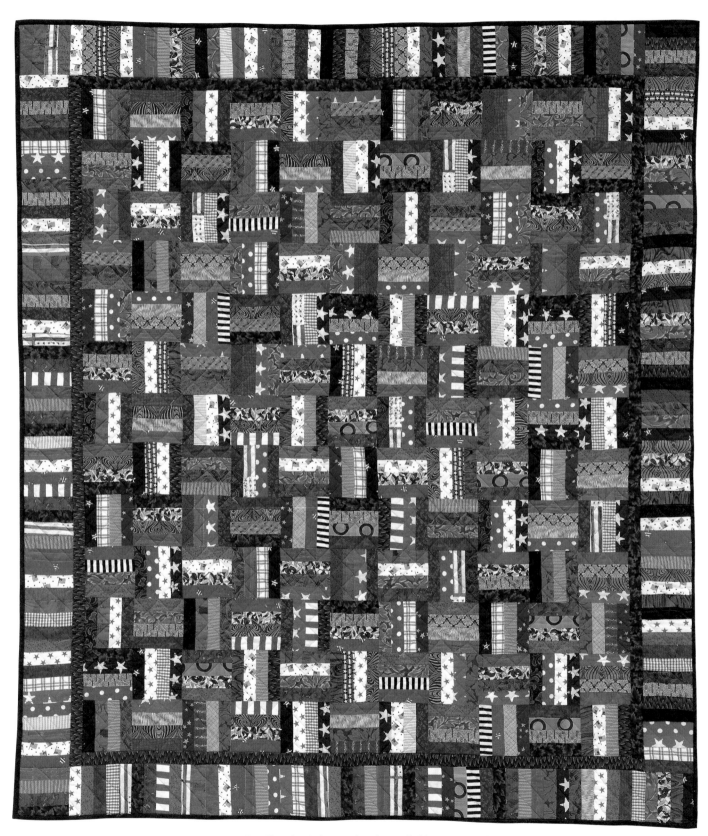

Pieced by Alex Anderson and machine-quilted by Patsi Hanseth.

RAIL FENCE AMERICANA

Hurrah for the red, white and blue! This vibrant quilt conjures up images of the Fourth of July, with flags flying, bands playing and fireworks illuminating the sky.

The quilt began with a collection of blue fabrics, always the easiest color to organize. I couldn't find enough good reds, and when I called "help" to Alex, she came to the rescue with a diverse array. My favorite is the one with large white stars on bright red ground, the perfect visual image for this quilt. Stripes, plaids, checks, polka dots and leaves are all present, and I especially like the fact that each of these images is shown in various sizes. For example, there's a very narrow red and white stripe and a wide version of the same design.

White fabrics are also very effective, both for contrast and for visual impact. Red stars sparkle on white ground; small flags flutter against a light sky. White provides necessary light contrast when placed against the medium and dark values of reds and blues. There is also a good balance between bright and dull fabrics.

Notice how the border shades from red to red-white-blue to blue, and how Alex cleverly places a red inner border next to the mostly blue pieced border, and a blue inner border next to the mostly red pieced border. This added touch contributes to making *Rail Fence Americana* a special quilt.

Of course you can make this quilt in any color combination you choose, and you can also experiment with color and fabric placement to obtain different design effects. You might decide this pattern is the best candidate for making a dent in your scrap bag or for making some gift quilts. I like the combination of nostalgia with a feeling of durability that contributes to the timeless appeal of this quilt. So does my nineteen-year-old son—he wants it!

HINT FROM MARY: *Variety of visual texture lends spice to your quilts.* Because the quilt shown here has a repetitive shape, you have the opportunity to indulge in a wide variety of fabric designs in order to maintain visual interest. The visual texture of a fabric refers to how the fabric looks: to the design elements and how they are arranged on the fabric.

When you are collecting lights, mediums and darks for this quilt, you can enlarge the scope of each category. For example, instead of just one or two medium blues, you can vary the visual texture by using several medium blues. The color remains the same; the value is still medium, but you have introduced a variety of textures into the medium blues. Alex has done this with all three colors and the result is a high degree of visual impact.

Of course, if you are interested in obtaining the opposite effect, you can use your knowledge of visual texture to create a feeling of repose by reducing the variety of prints. Experiment with your fabrics, laying them out in different arrangements, before making final decisions about how much visual texture contrast you wish to employ.

SIZE

This quilt measures 54" x 62", including a 1" inner border that is red on two sides and blue on two sides, and a 4" pieced outer border.

TEMPLATES

You will need one template: **RF 1**.

YARDAGE

RED, WHITE, BLUE PRINTS	3½ to 4 yards total (about 25 prints)
RED BORDER	¼ yard
BLUE BORDER	¼ yard
BACKING	4 yards (2 lengths)
BINDING	⅔ yard

CUTTING

RED, WHITE BLUE PRINTS	**RF 1** 788 (572 for blocks; 216 for border)
RED BORDER	Cut 3 strips 1½" wide on the *crosswise* grain of the fabric.
BLUE BORDER	Cut 3 strips 1½" wide on the *crosswise* grain of the fabric.

CONSTRUCTION

1. Each 4" block is made from 4 **RF 1** rectangles. Piece 143 blocks.

2. Assemble in 13 rows of 11 blocks each.

3. Sew the completed rows together.

4. Measure the length of the quilt top to determine the measurements for the side inner border strips.

5. Trim one red and one blue border strip to the correct side measurement and sew to each side.

6. Measure the width of the quilt top, and trim one red and one blue border strip to the top and bottom measurement and attach. Refer to the photograph to see where the red and blue are placed.

7. To piece the outer border: sew two rows of 54 rectangles. Attach one completed row to each side.

8. Repeat, piecing two rows of 54 rectangles. Notice in the photograph of the quilt how the blocks are rotated or turned. Look carefully at the beginning of each row to see how the placement of the blocks alternates. One row begins with the block placed horizontally, and the next row begins with the block rotated so that it is placed vertically. Refer to the photograph to arrange your placement, and then sew finished strips to the top and bottom.

9. Prepare the finished quilt top for quilting. Press carefully, taking care not to distort the edges. Layer and baste the quilt top, batting and backing. Hand or machine quilt.

10. Finish the edges by attaching the binding.

QUILTING SUGGESTIONS

This is an active quilt, so I like the idea of very simple quilting. Diagonal lines are used to cross-hatch the entire surface, spaced out about an inch and a quarter apart. The quilting design extends unbroken through the borders to the edges of the quilt.

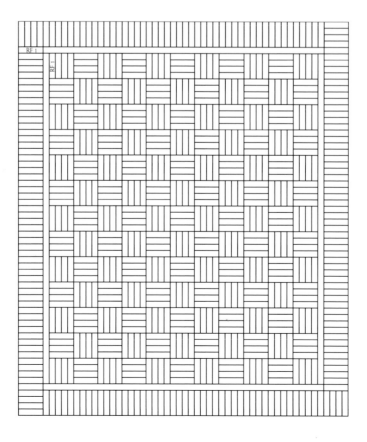

BRICK TRIP AROUND THE WORLD

Alex chose the fabrics for this quilt, and I think they're gorgeous. The colors of the black print in the very center establish the harmony of soft lavenders and greens radiating out to deeper values of the same colors. Only ten fabrics are used to create pleasing contrasts: four lavenders, three greens and three blacks. In addition to the black floral, there is a dark fabric that shades from blue to navy, and a strong black fabric with white swirls. Notice that one of the lavenders appears in a medium value as a surprise in the corners of the quilt.

Upon seeing this quilt, a friend remarked that the center reminded him of the Ojo de Dios or Eye of God design prevalent in the southwest. The center rectangle is distinct from the fabrics surrounding it because of the light-dark contrast. In fact, the entire quilt is an excellent example of light and dark placement in order to establish design and produce contrast.

HINT FROM MARY: *Value establishes pattern and contrast.* Value refers to the amount of light or dark in a color or fabric. Look carefully at the fabrics you have chosen to work with; sort them into lights, mediums and darks. Is one of these categories missing? Do you have seven mediums, two lights and one dark? Do you have ten mediums?

Go back to your fabric stash or to the quilt shop until you are satisfied that you are working with fabrics that establish value contrasts, which in turn establish the design of the *Brick Trip Around The World.* The trip to find a variety of values in fabrics is guaranteed to be enjoyable!

SIZE

This quilt measures 58" x 60".

TEMPLATES

You need 2 templates: **BT 1**, **BT 2**.

YARDAGE

Nine fabrics are used to make the quilt shown here. Refer to the diagram of the quilt for placement of the fabrics.

#1	½ yard
#2	⅔ yard
#3	⅔ yard
#4	¾ yard
#5	½ yard
#6	⅔ yard
#7	⅝ yard
#8	¼ yard
#9	¼ yard
BACKING	3½ yards (2 lengths)
BINDING	⅔ yard

CUTTING

The chart below indicates how many pieces to cut from each template.

Fabric	Template BT 2	Template BT 1
#1, #5	36 each	4 each
#2, #3, #6	68 each	4 each
#4	69	4
#7	52	4
#8	16	
#9	8	

Pieced by Alex Anderson and machine-quilted by Rhondi Hindman.

QUICK CUTTING

1. Cut all strips 2½" wide on the *crosswise* grain of the fabric according to the following chart.

Fabric	Number of Strips
#1, #5	5 each
#2, #3, #6	8 each
#4	9
#7	7
#8	2
#9	1

2. Take 4 rectangles from each of these fabrics: #1, #2, #3, #4, #5, #6 and #7. Cut them into 2½" squares. At this point, it's a good idea to remove all the leftover pieces from your work area so that you don't use them accidentally in place of the small squares.

CONSTRUCTION

The *Brick Trip Around The World* is constructed from 29 horizontal rows of 15 rectangles each. A few rectangles have been cut in half, in order to fill in the sides.

1. Sew in horizontal rows, following the diagram.

2. Remember that every other row begins and ends with a square.

3. Assemble the finished rows, referring to the diagram for placement.

4. Prepare the finished quilt top for quilting. Press carefully, taking care not to distort the edges. Layer and baste the quilt top, batting and backing. Hand or machine quilt.

5. Finish the edges by attaching the binding.

QUILTING SUGGESTIONS

First, the quilting design follows each row horizontally from side to side. Then, beginning in the center, each quadrant is quilted on the diagonal, extending to the outside corners. Diagonal stitching lines are one and one-half inches apart. This design produces a pleasing pattern of triangles on the surface of the quilt, a nice contrast with the rectangles.

1	2	3	4	5	6	7	2	7	6	5	4	3	2	1	
1	2	3	4	5	6	7	2	2	7	6	5	4	3	2	1
2	3	4	5	6	7	2	3	2	7	6	5	4	3	2	
2	3	4	5	6	7	2	3	3	2	7	6	5	4	3	2
3	4	5	6	7	2	3	4	3	2	7	6	5	4	3	
3	4	5	6	7	2	3	4	4	3	2	7	6	5	4	3
4	5	6	7	2	3	4	1	4	3	2	7	6	5	4	
4	5	6	7	2	3	4	1	1	4	3	2	7	6	5	4
5	6	7	2	3	4	1	6	1	4	3	2	7	6	5	
5	6	7	2	3	4	1	6	6	1	4	3	2	7	6	5
6	7	2	3	4	1	6	8	6	1	4	3	2	7	6	
6	7	2	3	4	1	6	8	8	6	1	4	3	2	7	6
7	2	3	4	1	6	8	9	8	6	1	4	3	2	7	
7	2	3	4	1	6	8	9	9	8	6	1	4	3	2	7
2	3	4	1	6	8	9	4	9	8	6	1	4	3	2	
7	2	3	4	1	6	8	9	9	8	6	1	4	3	2	7
7	2	3	4	1	6	8	9	8	6	1	4	3	2	7	
6	7	2	3	4	1	6	8	8	6	1	4	3	2	7	6
6	7	2	3	4	1	6	8	6	1	4	3	2	7	6	
5	6	7	2	3	4	1	6	6	1	4	3	2	7	6	5
5	6	7	2	3	4	1	6	1	4	3	2	7	6	5	
4	5	6	7	2	3	4	1	1	4	3	2	7	6	5	4
4	5	6	7	2	3	4	1	4	3	2	7	6	5	4	
3	4	5	6	7	2	3	4	4	3	2	7	6	5	4	3
3	4	5	6	7	2	3	4	3	2	7	6	5	4	3	
2	3	4	5	6	7	2	3	3	2	7	6	5	4	3	2
2	3	4	5	6	7	2	3	2	7	6	5	4	3	2	
1	2	3	4	5	6	7	2	2	7	6	5	4	3	2	1
1	2	3	4	5	6	7	2	7	6	5	4	3	2	1	

1. Purple
2. Black print
3. Blue-gray
4. Black and pink
5. Green
6. Light blue-green
7. Light green
8. Pink and gray check
9. Lavender

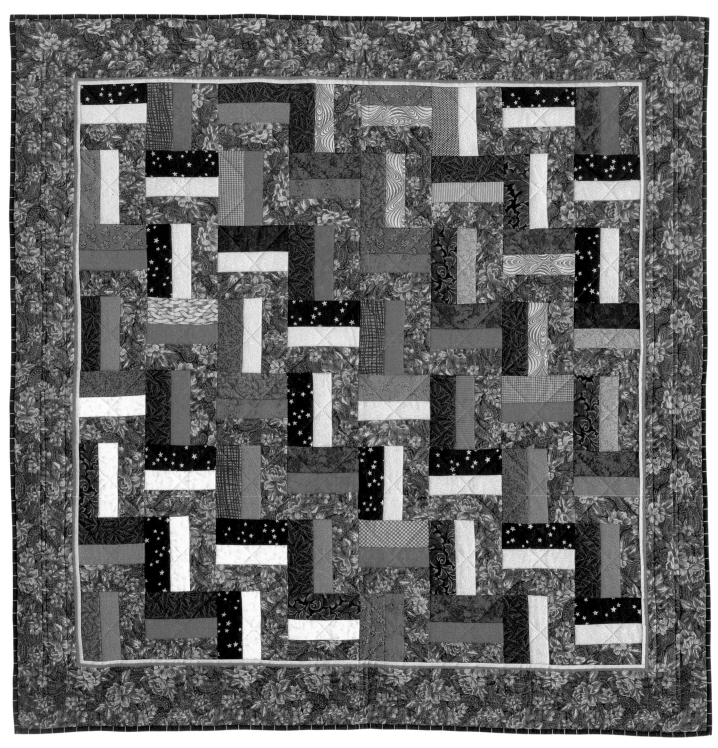

Pieced by Alex Anderson and machine-quilted by Patsi Hanseth.

ROMAN SQUARE

Here's a quilt that suggests the brilliant oranges, yellows and golds of fall foliage, viewed against a night sky. Vivid oranges and yellows are contrasted against subdued beiges, browns, grays, blacks and greens. This quilt began when I came across the lovely large floral that zigzags down through the squares and encloses the design with a wide border.

After choosing this fabric, it was easy to decide on the remaining colors. In this instance, the supporting fabric colors are all found in the main fabric. This is a potentially dangerous situation, because we tend to over-match when we go to the quilt store with a piece of fabric that is destined to be the inspiration for a quilt. But there are contrasts we can employ to avoid this pitfall. For this particular quilt, I decided to use the contrast of *intensity*, which refers to the brightness or dullness of a color.

If you look carefully at the floral fabric, you can see that the oranges and golds are subdued or dull. This is known as low intensity. By selecting brilliant oranges and bright yellows, which are high-intensity colors, I found the contrast the floral fabric needs in order to function effectively. Notice how well the browns and grays are getting along together in this quilt, and also the sparkling surprise of the black and gold star fabric.

You can make this quilt in any color harmony you like; I hope you'll experiment with the bright-dull contrast. You can also experiment with placement of colors and fabrics to create different design effects. The *Roman Square* is an easy quilt to make, so you can afford to put some time into exploring various options before sitting down at your sewing machine or threading your needle.

HINT FROM MARY: *Contrast may be achieved through the use of bright and dull fabrics.* When you shop for fabrics, be aware that the majority of prints are of medium to low intensity; they tend to have a grayed or dull look. Look for the high intensity or bright prints that provide a contrast. Brights are essential for accents in an otherwise dull quilt. Try using them in small amounts and I think you'll find the solution to the perennial lament: "I wish I could put some pizzazz into my quilts!"

SIZE

This quilt measures 54" square. Dimensions include a 5" border (¼" gold, ¼" orange, and 4½" floral).

TEMPLATES

You will need two templates: **RS 1**, **RS 2**.

YARDAGE

YELLOW	⅓ yard
ORANGE	⅓ yard
TAN	¼ yard
ASSORTED PRINTS	⅞ to 1 yard total
FLORAL	1⅝ yards (includes border)
ORANGE BORDER	¼ yard
GOLD BORDER	¼ yard
BACKING	3¼ yards (2 lengths)
BINDING	⅔ yard (crossgrain)

CUTTING

YELLOW	**RS 1**	27 rectangles
ORANGE	**RS 1**	28 rectangles
TAN	**RS 1**	9 rectangles
ASSORTED PRINTS	**RS 2**	64 rectangles
FLORAL		Cut 4 rows of border strips 5" wide on the *lengthwise* grain of the fabric.
	RS 2	64 (cut from remaining fabric)
ORANGE BORDER		Cut 6 rows of border strips 3/4" wide on the *crosswise* grain of the fabric.
GOLD BORDER		Cut 6 rows of border strips 3/4" wide on the *crosswise* grain of the fabric.

QUICK CUTTING

YELLOW — Cut 4 rows of strips 2" wide on the *crosswise* grain of the fabric. Cut into 6" rectangles for a total of 27.

ORANGE — Follow instructions for YELLOW, for a total of 28.

TAN — Cut 2 rows of strips 2" wide on the *crosswise* grain of the fabric. Cut into 6" rectangles for a total of 9.

ASSORTED PRINTS — Stack scraps and cut 2½" x 6" rectangles for a total of 64.

FLORAL — Cut 4 rows of border strips 5" wide on the *lengthwise* grain of the fabric.

From the remaining fabric, cut 8 rows of strips 6" wide. Cut into 2½" rectangles for a total of 64.

RED BORDER — Cut 6 rows of border strips ¾" wide on the *crosswise* grain of the fabric.

GOLD BORDER — Cut 6 rows of border strips ¾" wide on the *crosswise* grain of the fabric.

CONSTRUCTION

The *Roman Square* quilt shown here is made from 64 blocks set side by side. There are 8 rows of 8 blocks each. The design is created by rotating the blocks. Each 5½" block has three rectangles: one floral (**RS 2**), one assorted print (**RS 2**), and one yellow, tan or orange rectangle (**RS 1**). The rectangles are arranged **RS 2** - **RS 1** - **RS 2**. You can experiment with color and fabric placement, as well as with rotation, to create other effects from this design.

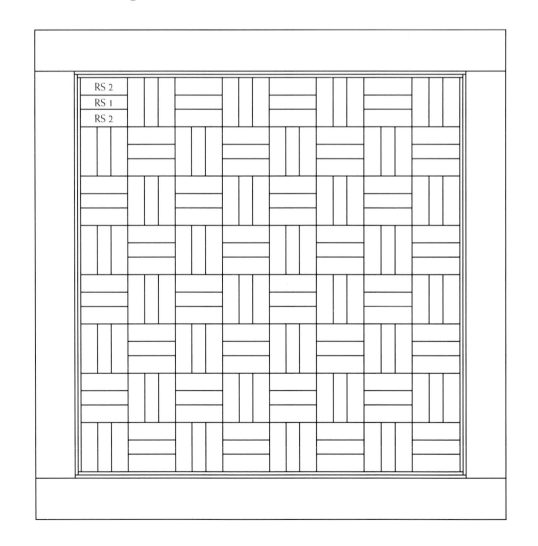

1. Piece 64 blocks.

2. Arrange the completed blocks as shown, or in a different rotation that pleases you. Sew in rows of 8 blocks.

3. Attach the rows.

4. Measure the length to determine the measurement for trimming the side border strips to fit.

5. Trim the two gold side borders and sew them to the sides.

6. Measure the width of the quilt top. Trim the two gold top and bottom borders and attach.

7. Follow the same sequence for the orange border, measuring the length and width to determine measurements for trimming. Measure, trim and sew the sides first, and then the top and bottom.

8. Repeat in the same order for the outer floral border.

9. Prepare the finished quilt top for quilting. Press carefully, taking care not to distort the edges. Layer and baste the quilt top, batting and backing. Hand or machine quilt.

10. Finish the edges by attaching the binding.

QUILTING SUGGESTIONS

Each rectangle is quilted in the ditch or seam. Squares composed of three rectangles are then quilted on the diagonal from corner to corner. The resulting criss-cross effect produces a nice contrast with the linear progression of the rectangles. The borders are simply quilted with straight lines of stitching.

Pieced by Alex Anderson and machine-quilted by Rhondi Hindman.

BRICK ZIGZAG

Pink, blue and green: popular colors, aren't they? But they can be difficult to use without coming down with a case of the blahs. Here's a quilt that avoids that dread result, along with some pointers for bringing pizazz to three pleasing colors used in combination. This quilt succeeds for several reasons, beginning with Alex's choice of the large floral with the bright rose set against green leaves and small blue flowers.

There are four values of red in this quilt: the light pink check deepens to a rose floral, which in turn darkens to burgundy leaves. Best of all is the spattered pink with stars, a fun addition of a non-traditional fabric. Next, let's look at the blues: light leaves are placed next to dark stalagmites, which are adjacent to medium blue crystals. Three greens, a light all-over small green floral, medium green geometric leaves, and dark green and black leaves supply contrasting greens. You can see that a variety of values and visual textures is the best prescription for turning the blahs into a triumph.

I've saved the best for last. The pale watercolor print that is blue in some places and green or pink in others is a special fabric. It functions as a superb light contrast in this quilt; it may also be employed in place of a solid fabric for background in other quilts. This fabric adds subtle visual interest through very soft coloration and muted design, and I think you should buy a good-sized piece if you see it (or a fabric similar to it). Don't save it! Try it out for different effects and then use it.

HINT FROM MARY: *Buy fabrics you need as well as those you like in order to establish a good working collection.* You can see from the HINTS in this book that value (light and dark), intensity (bright and dull) and visual texture (how the fabric looks) are important contrasts. Take this knowledge with you to the quilt shop and put it to work as you select fabrics to make a particular quilt, or as you choose additions to your fabric collection.

SIZE

This quilt measures 50" x 52½".

TEMPLATES

You need 4 templates: **ZZ 1**, **ZZ 2**, **ZZ 3**, **ZZ 4**.

YARDAGE

Twelve fabrics are used to make the quilt shown here:

#1	FLORAL	¾ yard
#2	DARK BLUE	½ yard
#3	1ST MEDIUM BLUE	⅓ yard
#4	2ND MEDIUM BLUE	⅓ yard
#5	PALE BLUE-PINK	⅝ yard
#6	DARK GREEN	⅓ yard
#7	MEDIUM GREEN	⅝ yard
#8	LIGHT GREEN	⅓ yard
#9	MEDIUM PINK	⅓ yard
#10	1ST MEDIUM ROSE	⅓ yard
#11	2ND MEDIUM ROSE	¼ yard
#12	DARK ROSE	⅓ yard
BACKING		3¼ yards (2 lengths)
BINDING		⅔ yard

CUTTING

To avoid confusion, it's a good idea to make labels for each stack as you cut.

Template	Cut this number	Use this fabric
ZZ 2	25	DARK BLUE #2
ZZ 2	25	DARK GREEN #6
ZZ 2	25	LIGHT GREEN #8

ZZ 2	50	PALE BLUE-PINK #5	
ZZ 2	25	DARK ROSE #12	
ZZ 2	30	FLORAL #1	
ZZ 2	29	MEDIUM PINK #9	
ZZ 1	50	DARK BLUE #2	
ZZ 3	25	1ST MEDIUM BLUE #3	
ZZ 3	25	2ND MEDIUM BLUE #4	
ZZ 3	50	MEDIUM GREEN #7	
ZZ 3	25	1ST MEDIUM ROSE #10	
ZZ 3	20	2ND MEDIUM ROSE #11	
ZZ 4	25	FLORAL #1	

QUICK CUTTING

Be sure to label each stack as you cut. Cut all strips 2½" wide on the *crossgrain* of the fabric.

Fabric	# of Strips	Rectangle Size	# of Rectangles
#1 FLORAL	4	5½"	25
	3	3½"	30
#2 DARK BLUE	3	3½"	25
	2	1½"	50
#3 MEDIUM BLUE	3	4½"	25
#4 MEDIUM BLUE	3	4½"	25
#5 PALE BLUE-PINK	6	3½"	50
#6 DARK GREEN	3	3½"	25
#7 MEDIUM GREEN	6	4½"	50
#8 LIGHT GREEN	3	3½"	25
#9 MEDIUM PINK	3	3½"	29
#10 MEDIUM ROSE	2	4½"	25
#11 MEDIUM ROSE	3	4½"	20
#12 DARK ROSE	3	3½"	25

CONSTRUCTION

The *Brick Zigzag* is made from 25 vertical rows of 2"-wide rectangles in 4 lengths.

1. Sew in vertical rows as shown in the diagram. You will notice that the rows are different lengths. Don't be alarmed! This allowance makes it easier to line up the rows, which are trimmed off later. Rows should measure 53" long.

2. To trim rows 1, 7, 13, 19 and 25: measure from the *top* of the row and trim to 53" long.

3. To trim rows 2, 6, 8, 12, 14, 18, 20 and 24: measure from the *bottom* of the row and trim to 53" long.

4. The remainder of the rows need trimming at both the top and the bottom. But before you do any cutting, first lay out all the rows according to the diagram. Check to make sure your layout corresponds to the diagram. The following trimming measurements may have to be altered slightly to ensure that the design flows properly across the quilt.

5. To trim rows 3, 5, 9, 11, 15, 17, 21 and 23: first trim one inch off the *top* of each one. Then measure from the *top* and trim to 53" long.

6. To trim rows 4, 10, 16 and 22: trim 3 inches off the *bottom* of each one. Then measure from the *bottom* and trim to 53" long.

7. Sew the rows together.

8. Prepare the finished quilt top for quilting. Press carefully, taking care not to distort the edges. Layer and baste the quilt top, batting and backing. Hand or machine quilt.

9. Finish the edges by attaching the binding.

QUILTING SUGGESTIONS

Simple quilting follows the lines of this pattern. First, each vertical row is quilted in the ditch or seam. Then horizontal rows are quilted, again in the ditch or seam at various intervals. The *Zigzag* is not a candidate for fancy quilting, so this simple treatment is very effective.

Trimming line

Trimming line

1. Black and pink floral
2. Dark blue
3. Medium blue #1
4. Medium blue # 2
5. Pale blue-pink
6. Dark green
7. Medium green
8. Light green
9. Medium pink
10. Medium rose #1
11. Medium rose #2
12. Dark rose

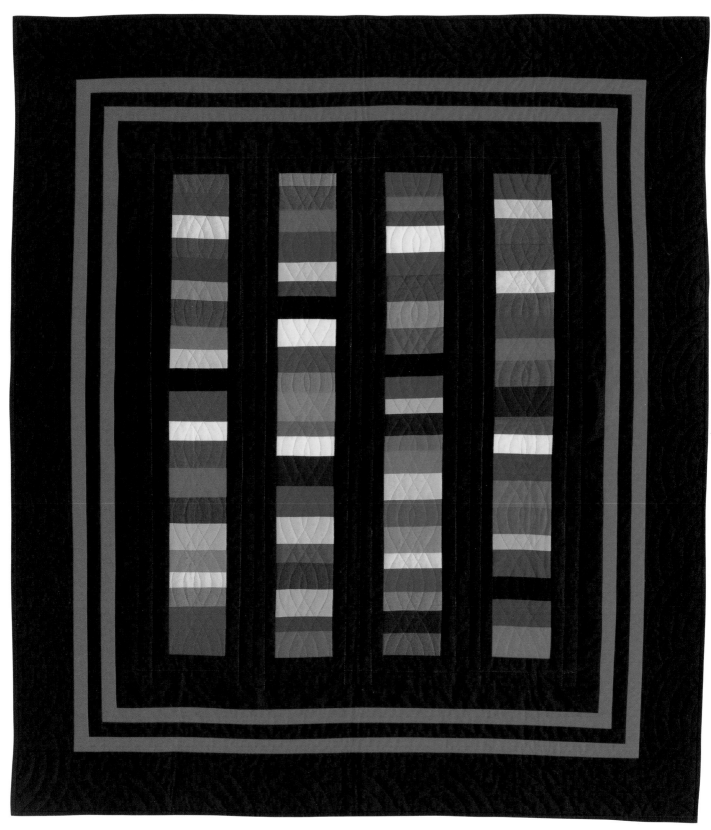

Pieced by Alex Anderson and machine-quilted by Rhondi Hindman.

CHINESE COINS

Here's a smashing rendition of an interesting design. This quilt illustrates what might happen when an Amish quiltmaker visits the Chinatown section of San Francisco! Alex has designed a clever melding of two color traditions and I think it comes off splendidly. The vivid colors are intensified against black, and the dull red inner borders serve to calm the composition as the eye recedes to the borders.

You may of course opt to use more traditional Amish colors, or you might even make another version using prints for the coins. I'd like to see a charm quilt version of this pattern, where each coin is a different fabric. Remember that the coins are assorted sizes, so be sure to retain the random feeling that contributes to the appeal of the design.

HINT FROM MARY: *Many influences from the past and present are inspirations for quilt designs.* This quilt definitely has an Amish flavor or influence because of black juxtaposed against vivid colors, and also because of the affinity of the design to the classic Amish bar quilt. The coin colors Alex has chosen have a fresh, contemporary look.

We can look beyond quilts to paintings, architecture, archaeology, nature, scenery, advertising, textiles, fashion, music, films and television to discover these influences and enlarge our perceptions of the color and design world. I think it is very important for a creative person to make this kind of investigation part of daily life. Inspiration is as close as looking out the kitchen window; it surrounds us when we take a walk or look at the products in the grocery store; it is abundant at the local library. Help yourself —it's free!

SIZE

This quilt measures 46½" by 51", including a 10" border arranged in this sequence: 3½" inner black, 1" red, 1" black, 1" red and 3½" outer black.

TEMPLATES

There are 6 rectangle templates: **CH 1**, **CH 2**, **CH 3**, **CH 4**, **CH 5**, **CH 6**.

YARDAGE

BLACK	1⅞ yards (includes borders and rectangles)
RED	1¼ yards (includes borders and rectangles)
ASSORTED COLORS	¾ to 1 yard. Over 20 solid-color fabrics are used in the quilt as shown, including several values of red, blue, green, purple, blue-violet, pink, teal, turquoise, fuchsia, aqua, yellow and gold. You need 80 to 90 rectangles.
BACKING	3 yards (2 lengths)
BINDING	⅔ yard

CUTTING

BLACK	Cut 4 strips 1½" wide and cut one strip 4" wide. Cut both strips on the *crossgrain* of the fabric for the narrow inner border.
	Cut 10 strips 4" wide on the *lengthwise* grain of the fabric.
	From the remainder, cut 5 to 7 rectangles of assorted sizes from the rectagle templates.
RED	Cut 8 strips 1½" wide on the *lengthwise* grain of the fabric.
	From the remainder, cut 8 to 10 rectangles of assorted sizes from the rectangle templates.
ASSORTED COLORS	Each column has 20 to 22 rectangles in various widths. You need a total of 80 to 90 rectangles. Use the 6 templates to produce a random arrangement of sizes.

CONSTRUCTION

1. Piece 4 columns of rectangles of assorted colors and widths. Columns should measure 31½" long. Experiment with your last few rectangles to reach this length.

2. Trim 5 black strips to 31½" long.

3. Sew the 4 pieced columns to the 5 black strips, alternating strips as shown in the photograph.

4. Trim 2 black strips to the required size and sew to top and bottom.

5. Measure the length to determine the measurements for the first red side border strips.

6. Trim 2 red side borders to the required measurement and attach to each side.

7. Measure the width and trim 2 red top and bottom borders and attach.

8. Take new measurements and trim two narrow black borders to fit the sides. Attach.

9. Measure, trim and sew the top and bottom black border.

10. Measure, trim and sew the second red border to each side.

11. Measure, trim and sew the second red border to the top and bottom.

12. Continue the same order of working for the wide black outer border. First measure the length; then trim and sew the sides. Repeat the process for the top and bottom.

13. Prepare the finished quilt top for quilting. Press carefully, taking care not to distort the edges. Layer and baste the quilt top, batting and backing. Hand or machine quilt.

14. Finish the edges by attaching the binding.

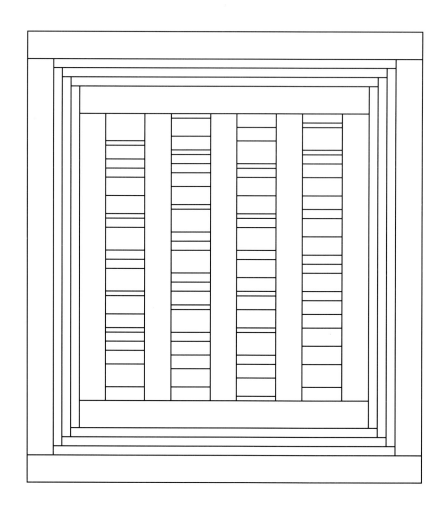

ADDITIONAL SIZE

To make the *Chinese Coins* in a larger size, measuring 76½" x 82", use the same 10" border. However, the 9 black strips and the 8 pieced strips each need to be 62" long. You will need the following yardage:

BLACK	4 yards
RED	2¼ yards
ASSORTED COLORS	2 to 2½ yards
BACKING	6 yards (2 lengths)
BINDING	1 yard

QUILTING SUGGESTIONS

Classic cables are used to enhance the pieced strips in this quilt. Straight lines run vertically through the alternating black strips. The narrow red and black borders are also quilted with vertical lines in the ditch or seam. The outer black border repeats the curved quilting motif with large quarter-circles touching the edges of the quilt.

Pieced by Kay Lettau and machine-quilted by Alex Anderson.

Special Project
LOG CABIN PLACE MAT

It's just not possible to have a rectangle book without a Log Cabin design! The idea was to keep the design entirely rectangles, so I've changed the familiar square chimney into a rectangle, a minor departure from the traditional version of this much-loved pattern. I think the Log Cabin is especially well-suited to the individual contrasts, and it's fun to concentrate on the colors and fabrics for just one block.

When I saw the gorgeous Japanese fabrics that had just arrived at The Quilt Patch, it was love at first sight. Yes indeed, this fabric is expensive! But I decided that a little would go a long way. Sometimes we have to splurge to create special design effects, and the effect I wanted, once I saw the fabric, was definitely Japanese. You can see this lovely fabric in both the center and in the side borders, where it shows to best advantage.

Collecting fabrics with a hint of a Japanese connection was great fun. While I knew that the blues, greens and yellows definitely needed the bright orange accent, I found that this fabric tended to overwhelm the others no matter where I placed it. Kay Lettau, who sewed the place mats, came up with the perfect solution: narrow orange strips on either side of the center and at the sides of the block. I think the effect is really handsome. I especially like the contrast of the Japanese fabric with the traditional American prints.

HINT FROM MARY: *Boost your creativity with some ethnic fabrics.* Quilting and fabric stores are brimming over with unusual fabrics from around the world. Look for batiks from southeast Asia, Provençal prints from France, tribal designs from Africa, intricate florals from England, vivid abstracts from Finland and lavish images from Japan. Buy a variety of designs in small amounts. Integrate ethnic fabrics into your collection, and use them sparingly to infuse your fabric combinations with visual zest.

SIZE

The place mat measures 12″ by 19″, including ½″ orange and 3″ Japanese floral borders on two sides.

TEMPLATES

You need 9 templates: **LC 1**, **LC2**, **LC 3**, **LC 4**, **LC 5**, **LC 6**, **LC 7**, **LC 8**, **LC 9**.

YARDAGE

Eight fabrics are used to make the *Log Cabin Place Mat* shown here. Yardage is given for making one set of four place mats. You can divide the amounts in half if you wish to make a pair. You may also use scraps for this project, perhaps designing a set in which each mat is made from different fabrics.

JAPANESE PRINT	½ yard
ORANGE	¼ yard
BLUE/GREEN/GOLD SYMBOLS	⅓ yard
YELLOW FLOWER	⅓ yard
MEDIUM BLUE FANS	⅓ yard
BROWN/YELLOW/GREEN BAMBOO	⅓ yard
DARK BLUE ROSES	⅓ yard
GREEN LEAVES	⅓ yard
BACKING	¾ yard
BINDING	⅓ yard

CUTTING

Use the following chart to cut pieces for one 12″ Log Cabin block. Cut all fabrics on the *crosswise* grain. After you lay out the block and you are sure the color and fabric placement are effective, then you can return to the chart and cut the additional pieces to make the set of four place mats.

Template	Size	Fabric	Cut
LC 1	2½″ x 3½″	Japanese floral	1
LC 2	1″ x 3½″	orange	2
LC 3	2″ x 3½″	blue/green/gold symbols	1
LC 4	2″ x 5″	blue/green/gold symbols	1
LC 4	2″ x 5″	yellow flower	1
LC 5	2″ x 6½″	yellow flower	1
LC 5	2″ x 6½″	medium blue fans	1
LC 6	2″ x 8″	medium blue fan	1
LC 6	2″ x 8″	brown/yellow/green bamboo	1
LC 7	2″ x 9½″	brown/yellow/green bamboo	1
LC 7	2″ x 9½″	dark blue roses	1
LC 8	2″ x 11″	dark blue roses	1
LC 8	2″ x 11″	green leaves	1
LC 8	2″ x 12½″	green leaves	1
ORANGE BORDER		Cut 2 strips 1″ wide on the *crossgrain* of the fabric.	
JAPANESE FLORAL BORDER		Cut 2 strips 3½″ wide by 12½″ long on the *lengthwise* grain of the fabric.	

QUICK CUTTING

On the *crosswise* grain of the fabric, cut orange strips 1″ wide. Cut Japanese floral strips 3½″ wide on the *lengthwise* grain of the fabric. Cut all the rest of the strips 2″ wide. Cut each strip to the required size according to the previous cutting chart.

CONSTRUCTION

1. Lay out all the pieces according to the diagram.

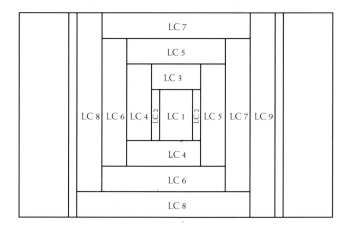

2. Follow the piecing order beginning with **LC 1** to make the Log Cabin block.

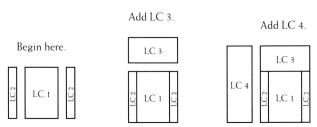

Begin here.   Add LC 3.   Add LC 4.

To assemble the 12″ Log Cabin block:
follow the numbers from LC 1 to LC 9.

3. Measure the width of the completed block and cut the orange strips to fit.

4. Sew orange inner border strips to each side.

5. Measure width again and cut the Japanese floral strips to fit.

6. Sew border strips to each side.

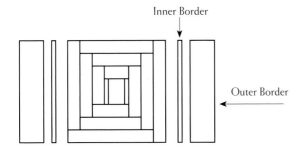

7. Prepare the finished place mat for quilting. Press carefully, taking care not to distort the edges. Layer and baste the place mat, batting and backing. Hand or machine quilt.

8. Finish the edges by attaching the binding.

QUILTING SUGGESTIONS

You may want to machine quilt your place mats for durability, although the portability of the small size invites hand quilting as well. While the quilting design is a simple outline of the rectangles, stitched in the ditch or seam, the use of gold metallic thread lends sparkle to the rich fabrics. Careful outline quilting of the Japanese floral in the two side borders produces an interesting effect, causing this fabric to appear pieced in certain areas.

TEACHING PLAN

The three books in this series present an excellent opportunity for quilting teachers and shop owners. Each individual book is suitable for a series of six lessons. Offering courses from all three books, *Squares, Triangles* and *Rectangles*, provides your students with consecutive lessons spanning three seasons of the year. Each student could expect to produce three quilt tops in this period of time (for example, from fall, winter and spring courses). You may wish to include a copy of the book in the tuition price for the course as a means of attracting students.

Preceding the course outline, you will find a structure for individual lessons. Using a lesson plan enables you, the teacher, to organize class time well; it also fosters learning because students respond enthusiastically to classes that are well presented. I hope you'll incorporate your own creative ideas into this plan.

SUGGESTED STRUCTURE FOR THE LESSONS

1. Introduction of lesson content
2. Presentation of supplies
3. Discussion of color and fabric theory
4. Demonstration of techniques
5. Group and individualized instruction
6. Evaluation and constructive criticism
7. Assignment of homework

After the course is underway, I recommend incorporating Show and Tell into every lesson. This is an enjoyable way to give recognition, inspiration and encouragement to students.

FORMAT FOR A SIX-LESSON COURSE

LESSON ONE — Emphasis on selection of quilt by each student OR assignment of quilt by the teacher if you prefer to teach one quilt to the entire class. Numbers 1, 2, 3 and 4 from the lesson structure are important parts of the first class.

LESSON TWO — Emphasis on in-store fabric-buying lesson, with selection of fabrics for each quilt. Planning color/fabric placement. Instruction for fabric preparation and cutting techniques. This is a busy class, but you'll still have time for numbers 1, 3, 5 and 7.

LESSON THREE — Emphasis on demonstration of techniques for machine or hand piecing. Include numbers 1 through 7.

LESSON FOUR — Emphasis on problem-solving and constructive criticism. Again, numbers 1 through 7 provide a good checklist.

LESSON FIVE — Emphasis on techniques for the addition of borders. Use the lesson structure as you introduce the new element of borders.

LESSON SIX — Emphasis on marking the top for quilting and making the binding. Summarize the main points of the previous lessons, and pay particular attention to numbers 5 and 6.

Schedule a Class Reunion so that you and your students may enjoy the finished quilts together!

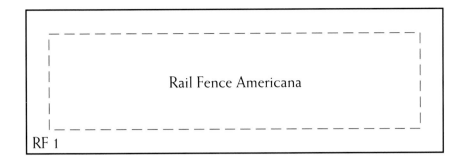

Rail Fence Americana

RF 1

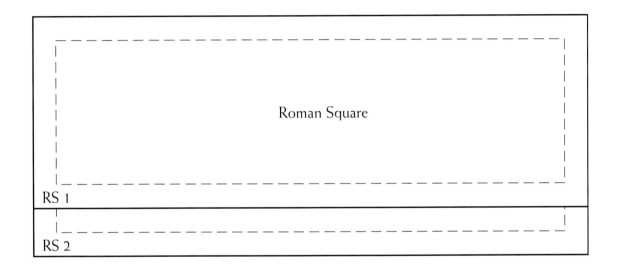

Roman Square

RS 1

RS 2

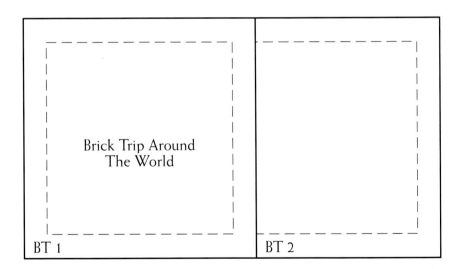

Brick Trip Around
The World

BT 1

BT 2

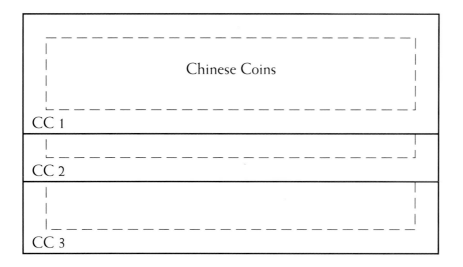

Chinese Coins

CC 1

CC 2

CC 3

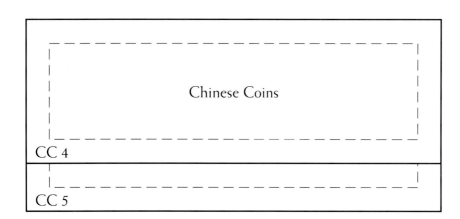

Chinese Coins

CC 4

CC 5

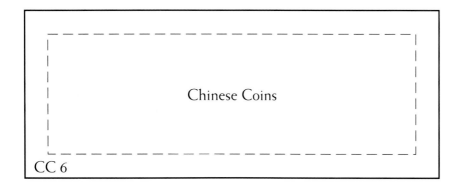

Chinese Coins

CC 6

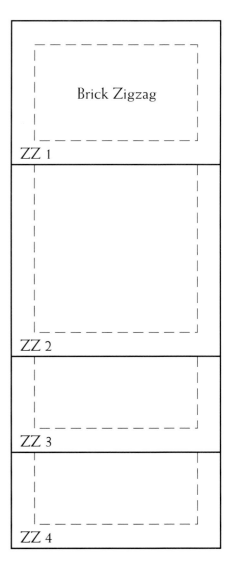

Brick Zigzag

ZZ 1

ZZ 2

ZZ 3

ZZ 4

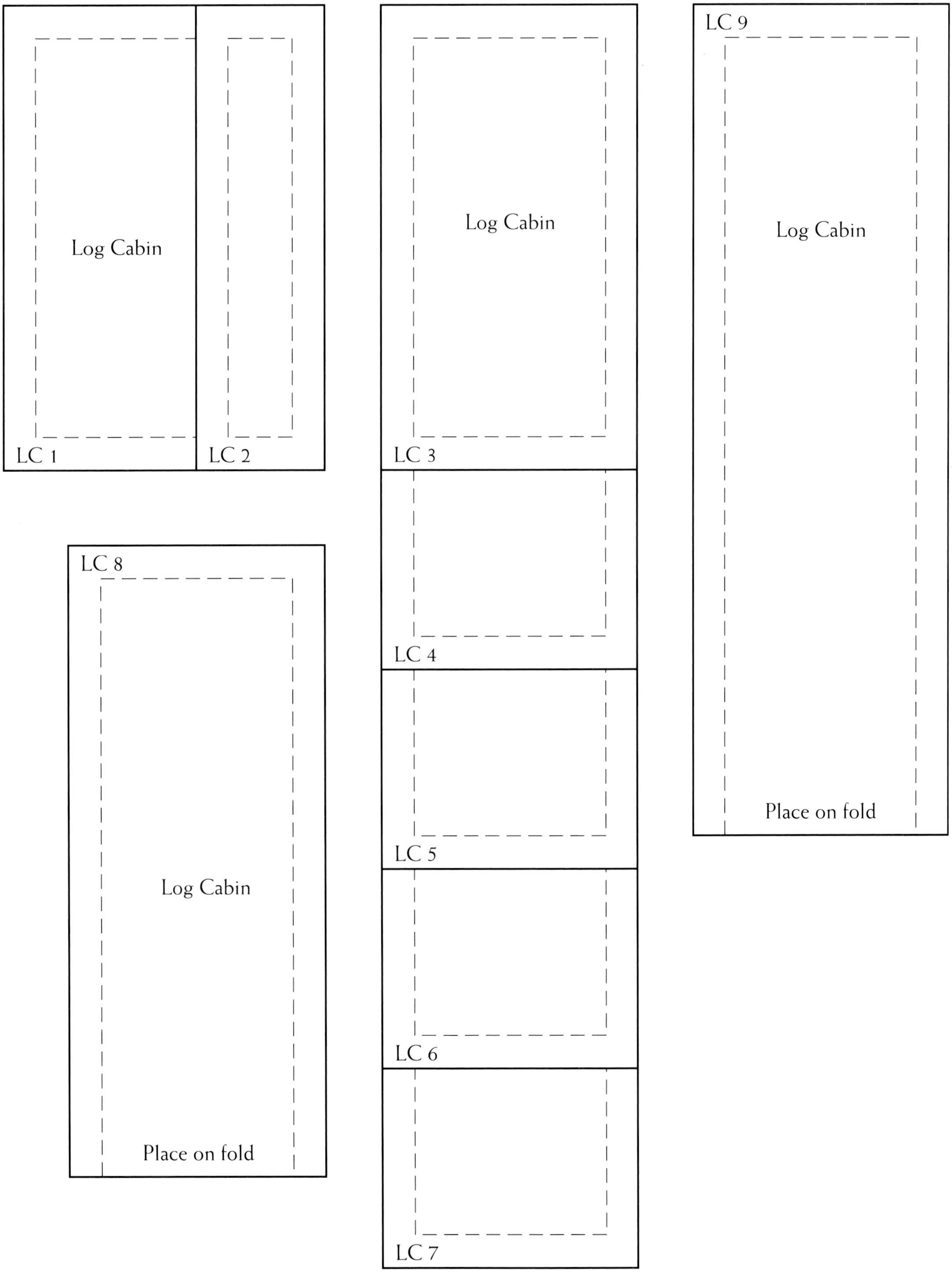

Log Cabin

LC 1

LC 2

Log Cabin

LC 3

LC 4

LC 5

LC 6

LC 7

LC 8

Log Cabin

Place on fold

LC 9

Log Cabin

Place on fold